Author Agent: longislandwriter@verizon.net

Copyright © Tim Roosevelt
ISBN-13: 978-1724499332
ISBN-10: 1724499335

Planes That Changed History

LOCKHEED SR-71 BLACKBIRD

Tim Roosevelt

Planes That Changed History - Lockheed SR-71 Blackbird

The SR-71 is an incredible airplane. It had a long career that began in the 1960s and ended in the 1990s. It served six presidents. It did things no other plane could. Its pilots were handpicked. Imagine flying one of the coolest looking planes in the world. And you can't even tell your friends about it. That's because your work is so secret.

The SR-71 seemed more like a spaceship than a plane. It looked like it was going fast even when it was standing still. The incredible plane flew at over 2,000 miles per hour. That's faster than a bullet. A car on the highway goes about a mile a minute. The SR-71 went faster than a mile every two seconds!

The SR-71 also flew very high. When you see small planes in the sky, they're about a mile or two above the ground. The SR-71 flew 16 miles above the ground. That's more than 80,000 feet high.

What does the world look like from up here? It looks like the photo above. It was taken from an SR-71 flying at around 83,000 feet. You're higher than any clouds. You're above thunderstorms or snowstorms. The sky is deep blue. You see the edge of space. You see the curve of the earth. It's almost like being an astronaut.

What was the SR-71 doing so many miles above the earth? It was collecting information for the military. That meant gathering information about naval bases, airfields, and other key places. The people at these places didn't want such information collected. When they saw the SR-71 in their skies, they tried to shoot it down.

But the SR-71 went about its mission. Its cameras and sensors gathered information on both sides of the plane. The SR-71 used cameras, radar, and other sensors. After the SR-71 landed, people on the ground removed the film and other information from the plane. They looked at it very closely.

When a plane flies as high and fast as the SR-71, it faces some problems. One of these problems is heat. Flying at 2,000 mph creates intense heat. Most planes are built out of a metal called aluminum. It is light and strong. But if it gets too hot, aluminum becomes weak and soft. This was not a problem in most planes because they flew at slower speeds.

The SR-71 was built with a special metal called titanium. Titanium is strong. It is light. And it stays strong even when it gets hot. Some places on the SR-71 became so hot they glowed cherry red.

The SR-71 had another way to deal with heat. The plane's skin became hotter than 500 degrees. The fuel on the inside of the plane "soaked up" that heat from the hot skin. This kept the plane cool. Wouldn't the fuel catch fire from absorbing this heat? We'll talk more about that later.

There was no escaping this intense heat. The plane's engineers had to protect the crew from it. They had to protect the plane's soft tires from it. It's funny to think of a plane that's red hot zooming through skies that are ice cold!

Another problem for the SR-71 was keeping pilots alive at 80,000 feet. Humans cannot survive at that altitude. Altitude means height above the earth.

Here on the ground, air pressure "pushes" on us from all sides. Without it, terrible things would happen to our bodies. SR-71 pilots wore pressure suits. These were like space suits used by astronauts. The suits kept pilots at the same pressure as being on the ground. The suits also provided oxygen. The air at 80,000 feet contains very little oxygen. We need oxygen in the air we breathe.

When people see an SR-71, they think, "What a cool plane!" But they might not realize that things are very harsh at this altitude. Sixteen miles above the earth, no human could survive in that cold and lifeless air.

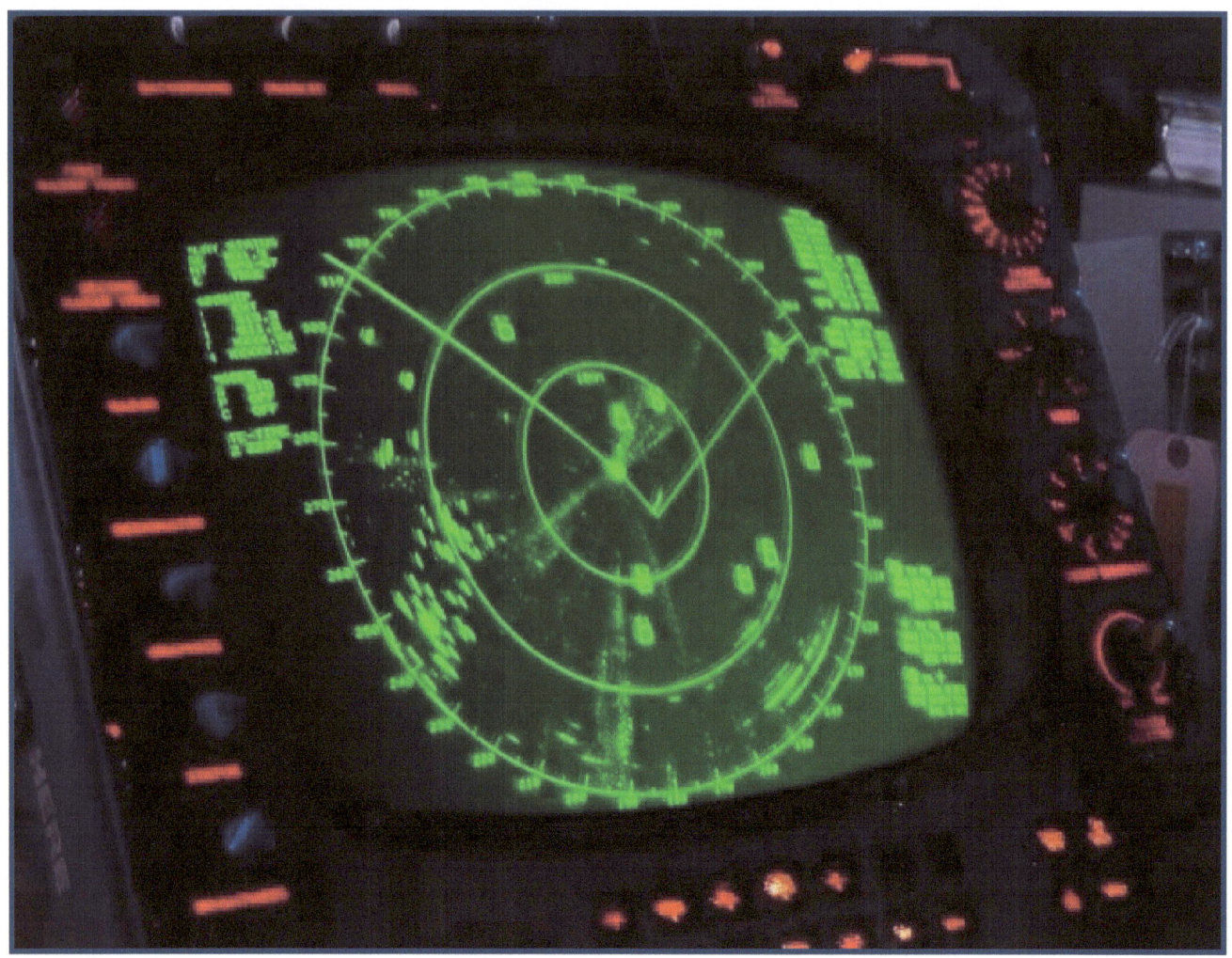

The SR-71 is retired. That means it no longer flies as a spy plane. But when SR-71s were in service, none of them were ever shot down. Many people tried to shoot the plane down. They knew it was up there. They fired missiles the size of telephone poles at the SR-71. They sent their best fighters up to shoot the plane down. But they never succeeded.

Why was this? The SR-71 was difficult to find with radar. The plane didn't show up well on radar screens like the one above. Radar is a tool that finds planes in the sky. But the SR-71 did a good job of "tricking" radar.

How did the SR-71 trick radar? Lots of thought went into creating its shape. It's easier to understand this when you look at the plane from the front.

The nose is wide and flat, with sharp edges. The tails are tilted inward. There are pointy cones at the front of each engine. They are called inlet spikes. These features made radar waves bounce off in all different directions. The plane was coated with a special black ferrite paint. This also helped trick radar. The paint absorbed the energy of the radar waves. Ferrite means the paint had iron in it.

The other reason why the SR-71 was difficult to shoot down was its incredible speed. It flew at three times the speed of sound. The speed of sound is the speed that sound waves travel through the air. When a plane flies at the speed of sound, we say it reaches Mach 1. That's about 741 mph.

Passenger jets like the Boeing 747 above do not reach Mach 1. Fast military jets reach Mach 1. Extremely fast military jets reach Mach 2, or twice the speed of sound. Very few planes reach Mach 2. The SR-71 reached Mach 3, or three times the speed of sound. No wonder it was so difficult to shoot down. It was gone before you knew it was there!

The reason for the SR-71's incredible speed was its engines. The plane used two Pratt & Whitney J58 gas turbine engines. These were extremely powerful. They are 15 feet long and about 4 feet in diameter. The photo above shows a J58 engine.

Let's get back to the matter of the SR-71 getting hot. A hot plane and jet fuel do not mix well. That's why a special fuel had to be made. The fuel would not catch fire in the hot SR-71. This special fuel was called JP-7. It even used bug spray in its formula! Nitrogen gas also helped here. A "blanket" of nitrogen gas covered the fuel in the plane's tanks. This kept air out. And a fire needs air to burn.

The SR-71 dripped fuel before taking off. The plane's body is basically a tube formed by six fuel tanks. The dripping was due to something known as "expansion." That means things get bigger when they become hot. They "expand."

The pieces that made up the SR-71 were made to expand as the plane grew hotter. If the pieces fit nice and snug when the plane was cool, what would happen when they became hot? They would expand. That would squeeze them together. And the pieces would break. So they fit loosely when the cool plane sat on the ground. That caused fuel to drip a little bit. But the pieces fit nice and snug when the plane got hot during flight. When that happened, the plane stopped leaking fuel.

The SR-71 could reach blazing speeds. Few planes even came close. But the SR-71 also burned huge amounts of fuel. Fuel is like gas for a car. A plane can only carry so much of it. Putting too much fuel on a plane makes it heavy. And that makes it slow.

The solution to this problem was to fuel the SR-71 in the air. Big tanker planes used a boom to pump fuel into the SR-71. A boom is a pipe. Fuel flowed from the tanker plane into the SR-71 through this pipe. The tanker plane was like a flying gas station. The one in the photo above is an Air Force KC-135 Stratotanker.

The SR-71 had a crew of two. The pilot sat in the front seat, facing the panel shown above. The pilot flew the plane. That meant controlling the plane's speed, altitude, and direction.

If you think of the SR-71 as a car, the pilot is the one at the steering wheel. The SR-71 was a difficult plane to fly. It wasn't simply sitting back and enjoying the view. At 2,000 mph, the smallest mistakes could be deadly. If an engine lost power, the whole plane shook with a bang. This was called an "unstart." It was as if an engine "burped." That may sound funny, but it was terrifying in real life for the crew.

The second crew member on an SR-71 was the Reconnaissance System Officer, or RSO. If you think of a car, the RSO is the one with a map who helps guide the driver. The RSO sat behind the pilot, facing the panel shown above.

Figuring out the plane's location is called navigation. The RSO handled navigation and made sure the plane was on the correct course. The RSO also operated the special cameras and sensors. The cameras took photos. The sensors picked up other kinds of information. The pilot and RSO worked tightly as a team.

Navigation is important for any plane. But in an SR-71, it was very important. At 2,000 mph, a small error in navigation would put the plane hundreds of miles off course very quickly.

It's hard to grasp just how fast the plane could go. This should help. An SR-71 flew from Los Angeles, California to Washington, D.C. in one hour and four minutes. An SR-71 flew from New York to London in one hour and fifty-five minutes. That means the plane crossed the Atlantic Ocean in less than two hours. Compare that with four days for an ocean liner like the *Queen Mary* to cross the Atlantic!

When you fly at such speeds, you see some strange things you don't see elsewhere. For instance, SR-71 pilots sometimes saw the sun rise more than once during a flight.

That was because the plane went toward the sun then away from the sun then toward the sun again at high speed. The plane seemed to "force" the sun to rise or set. When you fly that high above the earth, you see many, many stars. It would have been a beautiful sight. But it took lots of focus to fly the plane. So the crew couldn't really enjoy the view.

The SR-71 used stars to navigate. Hundreds of years ago, sailing ships also used stars to navigate. Ancient sailors knew that each star was supposed to be in a certain place in the night sky. That told them where they were on the vast ocean.

The SR-71 did the same thing with a special tool. It was called ANS. That stands for Astroinertial Guidance System. The ANS knew the positions of 61 stars. It would scan the sky for stars and pick out the ones it knew. Based on the positions of those stars, it knew where the SR-71 was. This system worked both day and night. The crews had a nickname for the ANS. They called it R2-D2, like the robot in the *Star Wars* movies.

When the SR-71 flew, it seemed to break all the rules. How could something go that fast? How could something take all that heat? Well, SR-71s needed lots of care. That's how.

That meant checking the plane for things that needed fixing. Everything on the SR-71 had to work perfectly. If they didn't, crews could get hurt. And there were many things to worry about. There were the complex engines. There were the control systems that made the inlet spikes move back as the plane went faster. There were the aluminum-coated tires inflated with high-pressure nitrogen. This was not your average plane. It had to deal with intense heat. Small things breaking could become disasters at 2,000 mph.

What happened to the SR-71? It was retired in the 1990s. That means it was taken out of service. It's hard to imagine that the Air Force would stop using such an amazing plane. So much work went into its creation.

The problem was that the SR-71 was expensive to operate. Tanker planes had to be in the air waiting for it. Those high speeds and temperatures are hard on a plane. They made things wear down fast. Planes like the U-2 could also take photos. The plane above is a U-2. It didn't reach Mach 3. It flew slower than 500 mph. But it was less expensive to operate than the SR-71.

There were other reasons why the SR-71 was retired. The plane first flew in the 1960s. At the time, it was the best way to get top secret photos from high above the earth. But today, satellites can do that.

You can see SR-71s in museums today. If you ever have the chance to get close to one, you can let your imagination wander. You could think about what it would be like to fly a top secret photo mission in the 1960s. You'd look down at the earth from 16 miles up. The metal around your engines would be red hot in places. And you'd know there wasn't a plane out there that could catch you!

THE END

How many SR-71s were built? There were 29 of the main versions built. These were the SR-71A models. They operated out of bases in Japan, England, and the United States. Two SR-71B trainer versions were built. One hybrid version, called the SR-71C, was built from an SR-71 and a YF-12A.

In 1976, an SR-71 set a world speed record when it flew at 2,193 mph, and a record for altitude when it reached 85,069 feet.

Serial Numbers: SR-71A: 61-7950 to 61-7955; 61-7958 to 61-7980; SR-71B: 61-7956 and 61-7957; SR-71C: 61-7981

Photo Credits: p. 1 U.S. Government, p. 2 Major Brian Shul USAF, p. 3 USAF, p. 4 CIA, p. 5 USAF, p. 6 NASA, p. 7 USN, p. 8 USAF, p. 9 Adrian Pingstone, p. 10 USAF, p. 11 CIA, p. 12 Dryden Flight Research Center, p. 13 USAF, p. 14 USAF, p. 15 USAF, p. 16 Dryden Flight Research Center, p. 17 NASA, p. 18 Dryden Flight Research Center, p. 19 Master Sgt. John Gordinier USAF, p. 20 NASA, cover Dryden Flight Research Center, opposite page NASA, back cover Dryden Flight Research Center

The appearance of U.S. Department of Defense (DoD) visual information does not imply or constitute DoD endorsement.

Planes That Changed History - Lockheed SR-71 Blackbird

Lockheed SR-71 Specifications

Length:	107 feet 5 inches
Span:	55 feet 7 inches
Height:	18 feet 6 inches
Maximum Speed:	2,193 mph
Maximum Test Altitude:	85,069 feet

Engines:	Two Pratt & Whitney J58s
Power:	32,500 lbf each
Maximum Weight:	140,000 lb (in-flight)
Crew:	two
Armament:	none
Unrefueled Range:	3,250 miles

Sources

Lockheed SR-71 Flight Manual

Development of the Lockheed SR-71 Blackbird, Clarence L. Johnson, Senior Advisor, Lockheed Corporation July 29, 1981, Approved for Release 09.27.2011 CIA RDP90B00170R000100050008-1

The U-2, OXCART, and the SR-71 - U.S. Aerial Espionage in the Cold War and Beyond, National Security Archive Electronic Briefing Book No. 74 Jeffrey T. Richelson, Editor October 16, 2002

Memorandum for the Secretary of the Air Force from Dick Cheney, Secretary of Defense; SR-71 Program Termination June 21, 1990

Memorandum for the Honorable Norman Hicks, House of Representatives, from William Lynn, Under Secretary of Defense; Explanation for Permanent Retirement SR-71 August 21, 1998

Central Intelligence Agency, Specifications SR-71

USAF, Specifications SR-71

www.ingramcontent.com/pod-product-compliance
Lightning Source LLC
Chambersburg PA
CBHW051831210526
45473CB00005B/1825